INDABA YEZINOMBOLO

THE NUMBER STORY

SMALL BOOK ONE

ENGLISH - ZULU

---

*Numbers Teach Children
Their Number Names*

written and illustrated by

MISS ANNA

Early Reader Edition of *The Number Story 1*
Bronze Medal Winner, 2016 Wishing Shelf Book Award

Library of Congress Control Number: 2018902040

Names: Miss Anna, author.
Title: Number story : numbers teach children their number names / Miss Anna.
Description: Portland, OR: Lumpy Publishing, 2018.
Identifiers: ISBN 978-1-945977-99-2 | LCCN 2018902040
Summary: The pictures and rhymes present stories which introduce numbers 0-10.
Subjects: LCSH Numeration—English--Zulu--Pictorial works--Juvenile literature. | BISAC JUVENILE NONFICTION /
Languages: English--Zulu
Classification: LCC QA141.3 .M57 2018 | DDC 513—dc23

Publisher: Lumpy Publishing
Website: www.missannabooks.com
Email: missanna@missannabooks.com

Paperback: ISBN 978-1-945977-99-2
Printed in the U.S.A.   1 3 5 7 9 10 8 6 4 2

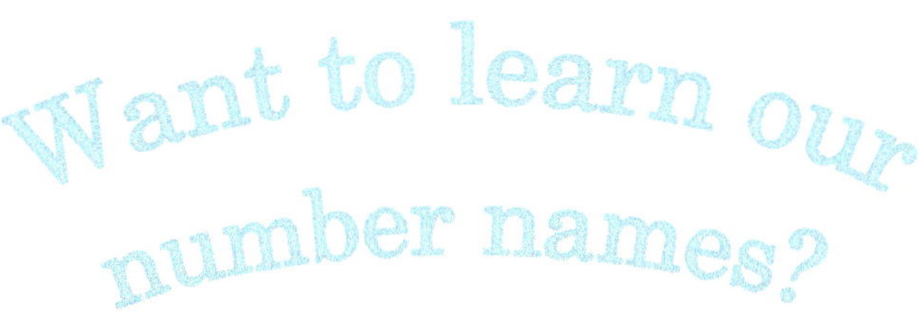

Ngabe ufuna ukufunda
amagama Ezinombolo zethu?

It is very easy and a lot of fun!

Kumalula kakhulu kanti futhi kumnandi!

Say-along our little jingle

Hlabelela nathi ingonyana encane!

starting from Number One!

Kusuka kwinombholo yokuqala!

1

ONE looks like my one finger.

KUNYE

Kubukeka kunjengowodwa umuno wami.

ONE!
KUNYE!

2

TWO trails a tail.

KUBILI

Uzama umsila.

A TAIL! UMSILA!

3

THREE   has bumps.

KUTHATHU

Kunamabhampi.

KUNAMABHAMPI!

4

FOUR carries a sail.

KUNE

Kuthwele iseyili.

ISEYILI!
Isikebhe esineSeyili!

FIVE   is a racing track.

ISIHLANU

Ngumvila wokuncintisana.

VROOM
VROOOM!

SIX   curves like a snail.

ISITHUPHA

Kuyijika njengomnenke.

A SNAIL! UMNENKE!

7

SEVEN has a sharp angle.

ISIKHOMBISA

Kunekona elicijile.

BE CAREFUL! IT'S SHARP!
QAPHELA! KUCIJILE!

8

EIGHT   is rollercoaster rails.

ISISHIYAGALOMBILI

Kuwujantshi we-*rollercoaster*.

YAY!
YIPPEE!

9

ISISHIYAGALOLUNYE

Ibhamuza endvukwini.

A BUBBLE!
IGWEBU!

10

TEN is an eye of a whale.

ISHUMI

Yiso le-whale.

HELLO!   SAWUBONA!

And
No

0

ZERO   is an empty pail.

AKUNANTO

Kokukhelela okungenalutho.

IT'S
EMPTY!
AKUNALUTHO!

Thank you for playing with us today.

We had a lot of fun too!

Siyabonga ukudlala nathi namhla-nje.

Sibe nesikhathi sokujabula nathi!

We are your Number friends,
Zero to Ten,
Who will be here for you~
Singabangani bakho bokuqala
Kusuka lakungenalutho khona kuya eshumini.
Sizohlala sikhona lapha senzele wena!

Bye-bye now!
See you again soon!
Sala kahle manje!
Sizokubona futhi masinya!

The Numbers are *SINGING* too!

To sing-a-long, look for Miss Anna Number Story
at your favorite music store like iTUNES.

MP3

Numbers 0-10
IDENTIFYING & COUNTING

Numbers 11-20
& Ordinals
first, second, third...

Numbers 0-100
& Place Values
ones, tens, hundreds...

About Clocks
& Telling Time
hours, minutes, seconds

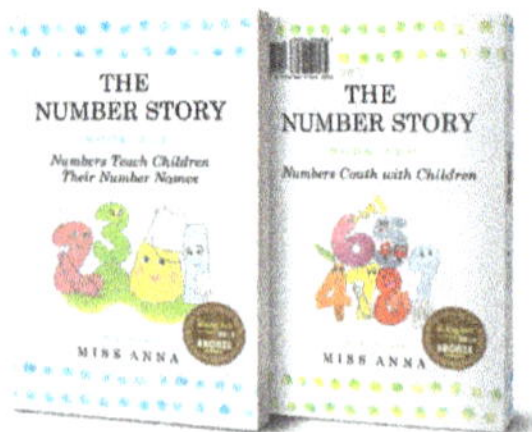

Number Story 1 & 2
isbn: 978-0-996216-48-7

Number Story 3 & 4
isbn: 978-1-945977-01-5

Number Story 5 & 6
isbn: 978-1-945977-06-0

Number Story 7 & 8
isbn: 978-1-949320-40-4

For more Miss Anna books to love,
visit us at

www.missannabooks.com

Numbers are working hard all over the world!
Come Travel the World with Us!

9 781945 977992